Parts of Me with Sage

Written By Veronica Cameron
Illustrated by Danh Tran Art

Hair
Eyes
Mouth
Chest
Hands
Legs

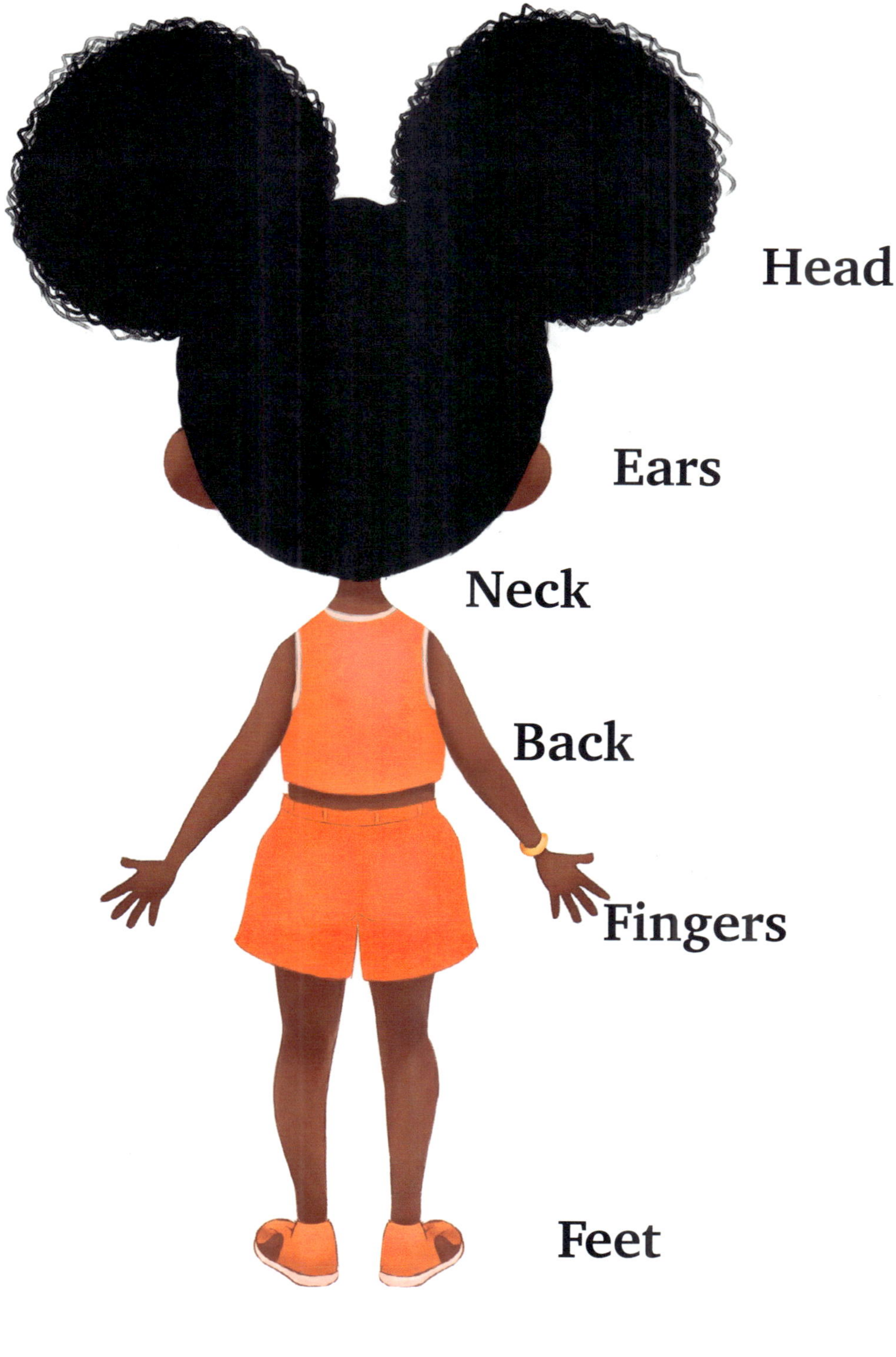
Head
Ears
Neck
Back
Fingers
Feet

This book belongs to...

....

Printed in the
United States of America
First Printing, 2021
ISBN 978-1-7343550-4-8
www.thecoolreeds.com

THE COOL REEDS®

FOR Zuri, Lena Simone, and Izrhi L.

LOVE YOUR BODY AND TREAT IT WELL.

Parts of Me with Sage

I love
my body,
yes, this is true!
Learning about it is
fun to do!

eye
knee
foot
leg
fingernails
mouth
hand
stomach
toes
body
shoulders
tongue
arm
nose
hair
thumb
ear
head
finger
lips

I love
my body,
yes, every part!
In fact, I know
a few by heart!

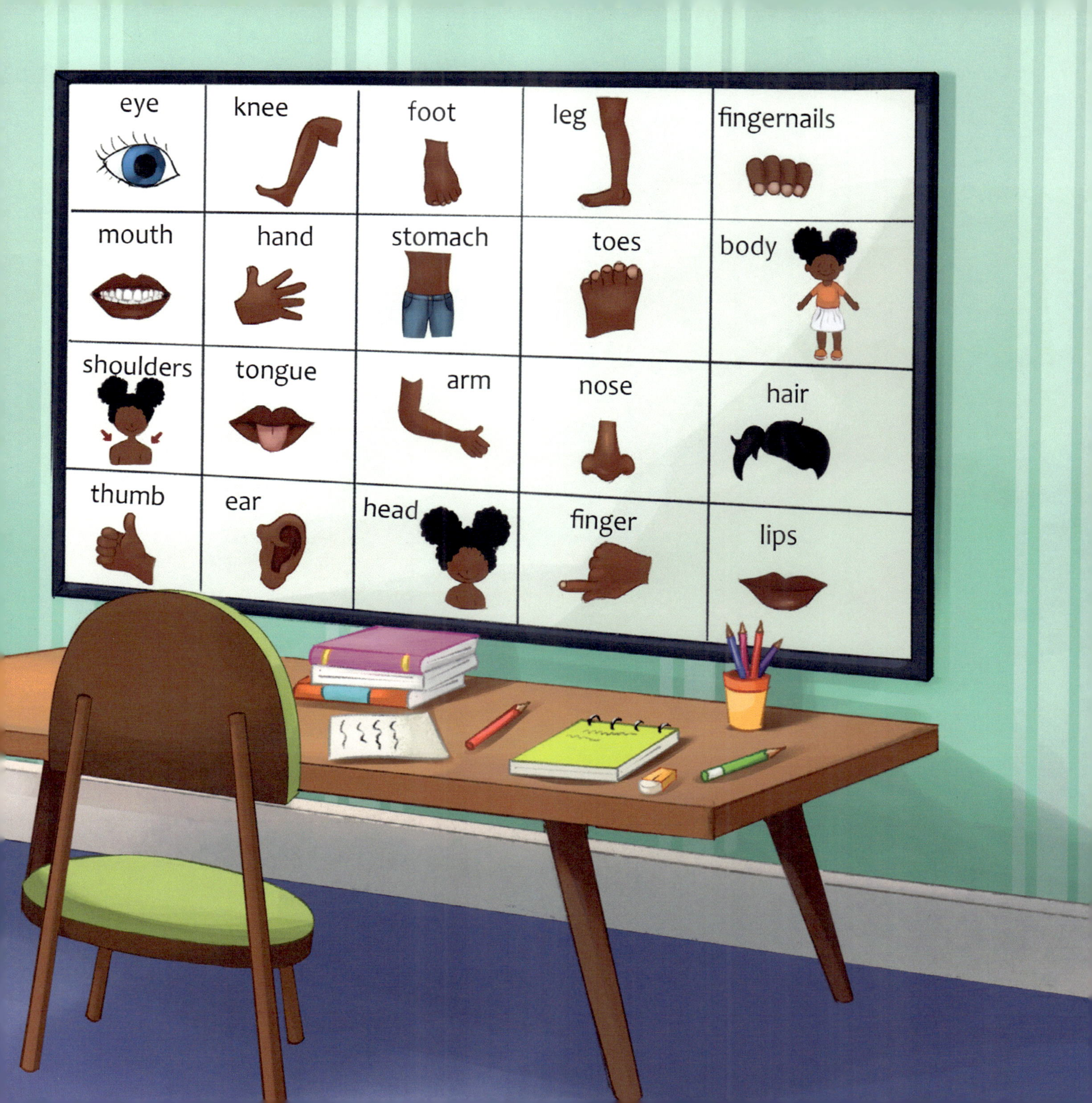
eye
knee
foot
leg
fingernails
mouth
hand
stomach
toes
body
shoulders
tongue
arm
nose
hair
thumb
ear
head
finger
lips

I have one head
with healthy hair,
I brush it and give it
lots of care!

I have two eyes and
two lovely ears,
to see, see, see,
and hear, hear, hear!

I have two nostrils
and one cute nose,
to follow smells
wherever they go!

I have one mouth
and lots of teeth,
to chew my food
and eat, eat, eat!

I have one neck
and two strong
shoulders,
they will get bigger
when I am older!

I have one chest
that holds my heart.
It pumps blood to my
body parts!

I have ten fingers
and two sturdy hands,
to build amazing
castles of sand!

I have two knees and
two powerful legs,
that help me run, and
jump, and bend!

I have two feet and
ten wiggly toes,
that take me where
I need to go!

I love my body,
yes, this is true!
Learning about it is
fun to do!

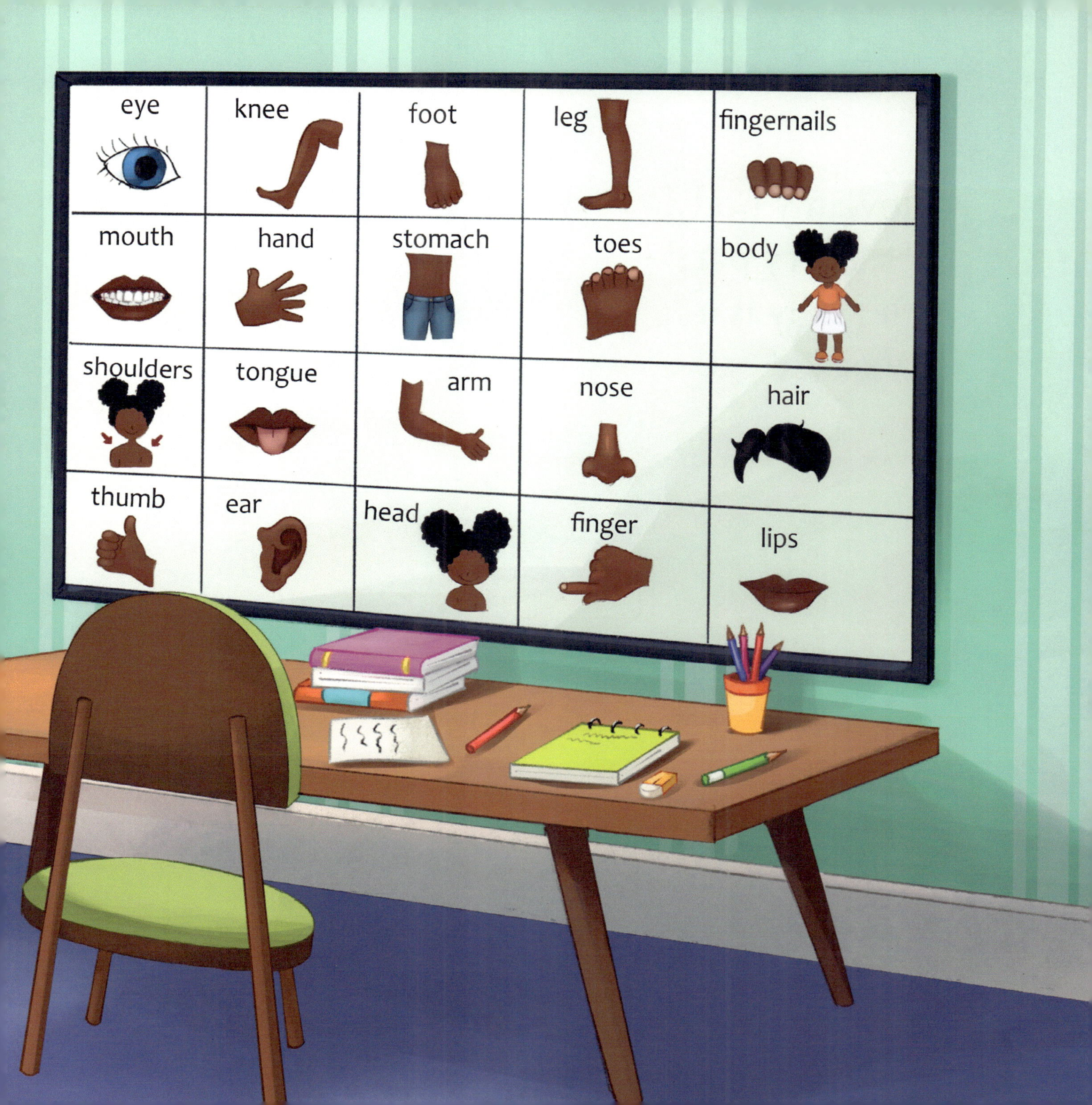
eye
knee
foot
leg
fingernails
mouth
hand
stomach
toes
body
shoulders
tongue
arm
nose
hair
thumb
ear
head
finger
lips

I love my body,
yes, every part!
It really is a
work of art!

eye
knee
foot
leg
fingernails
mouth
hand
stomach
toes
body
shoulders
tongue
arm
nose
hair
thumb
ear
head
finger
lips

Parts of Me with Sage
COLORING BOOK

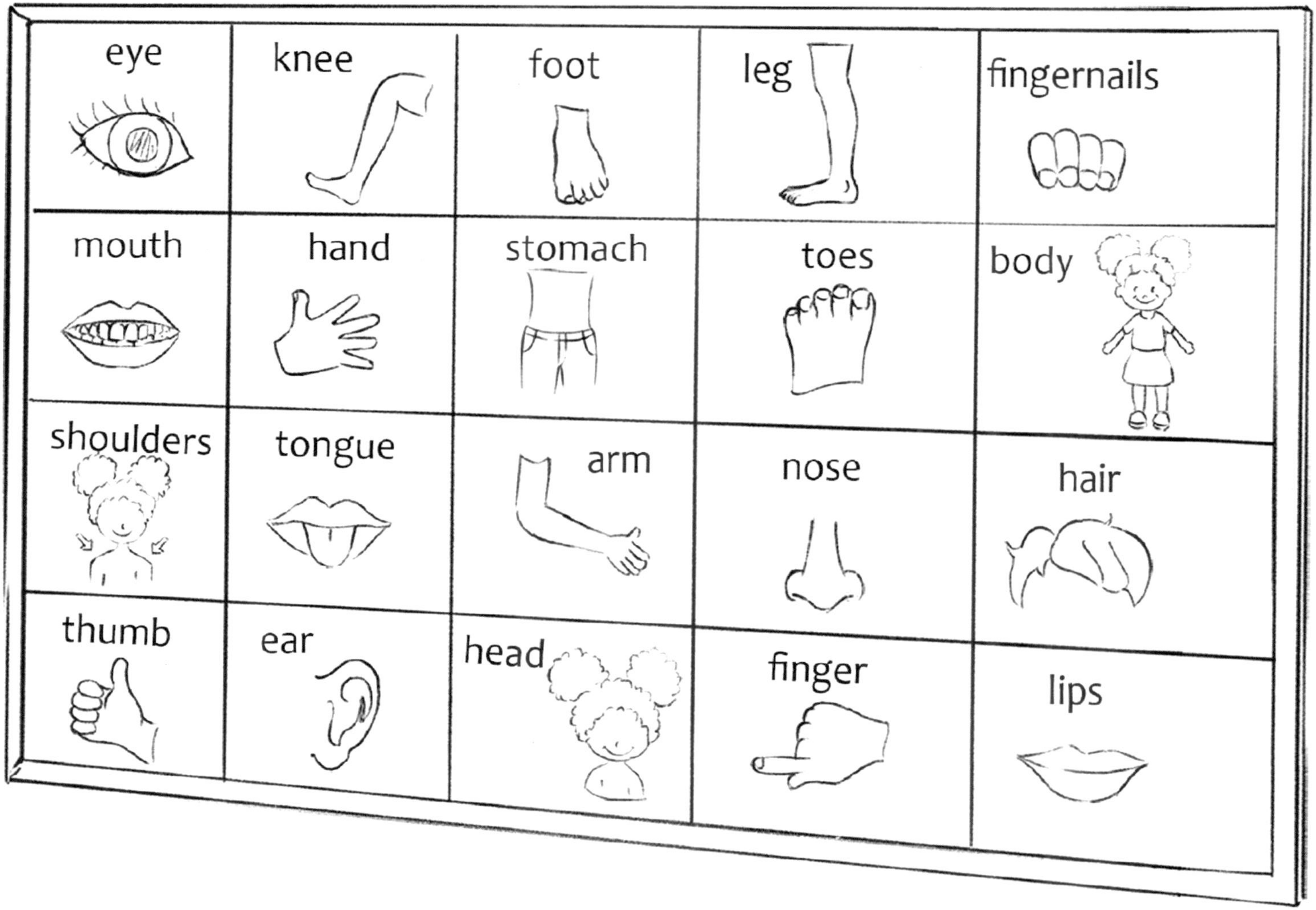
eye
knee
foot
leg
fingernails
mouth
hand
stomach
toes
body
shoulders
tongue
arm
nose
hair
thumb
ear
head
finger
lips

Hair

Eyes and ears

Nose

Mouth

Arms and shoulders

Hands

Legs and Knees

Feet

Made in the USA
Monee, IL
12 September 2026

62981954R00031